AUG 2020

T4-ADN-394

Caves
Julie Murray

Abdo Kids Junior
is an Imprint of Abdo Kids
abdobooks.com

Abdo ANIMAL HOMES Kids

abdobooks.com

Published by Abdo Kids, a division of ABDO, P.O. Box 398166, Minneapolis, Minnesota 55439. Copyright © 2020 by Abdo Consulting Group, Inc. International copyrights reserved in all countries. No part of this book may be reproduced in any form without written permission from the publisher. Abdo Kids Junior™ is a trademark and logo of Abdo Kids.

Printed in the United States of America, North Mankato, Minnesota.

052019

092019

THIS BOOK CONTAINS RECYCLED MATERIALS

Photo Credits: Getty Images, iStock, Shutterstock

Production Contributors: Teddy Borth, Jennie Forsberg, Grace Hansen

Design Contributors: Christina Doffing, Candice Keimig, Dorothy Toth

Library of Congress Control Number: 2018963309

Publisher's Cataloging-in-Publication Data

Names: Murray, Julie, author.
Title: Caves / by Julie Murray.
Description: Minneapolis, Minnesota : Abdo Kids, 2020 | Series: Animal homes | Includes online resources and index.
Identifiers: ISBN 9781532185229 (lib. bdg.) | ISBN 9781644941195 (pbk.) | ISBN 9781532186202 (ebook) | ISBN 9781532186691 (Read-to-me ebook)
Subjects: LCSH: Animal housing--Juvenile literature. | Caves--Juvenile literature. | Cave dwellings--Juvenile literature. | Animals--Habitations--Juvenile literature.
Classification: DDC 591.564--dc23

Table of Contents

Caves 4

What Lives in
a Cave? 22

Glossary. 23

Index 24

Abdo Kids Code. 24

Caves

Many animals live in caves.

A cave is an open space **underground**.

Most caves are dark. There is very little light.

A cave is a safe place to hide.

The lion is in a cave.

It keeps cool.

Bats live in caves. They hang upside down.

An **olm** lives in an underwater cave.

Glowworms live in caves. The light they make catches **prey**.

The bear is in a cave.

It will sleep all winter.

What Lives in a Cave?

blind salamanders

corn snakes

European cave spiders

Mexican free-tailed bats

Glossary

olm
a blind salamander only found in underwater caves.

prey
an animal being hunted, caught, or eaten by another animal.

underground
taking place beneath the earth's surface.

Index

bats 14

bear 20

characteristics 6, 8, 10

glowworms 18

light 8

lion 12

olm 16

sleep 20

underwater 16

Visit **abdokids.com** to access crafts, games, videos, and more!

Use Abdo Kids code
ACK5229
or scan this QR code!